ISBN 978-1-330-02997-8
PIBN 10008166

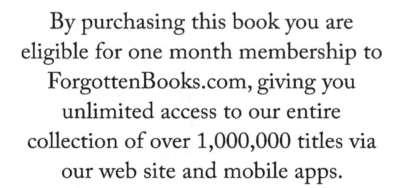

English
Français
Deutsche
Italiano
Español
Português

www.forgottenbooks.com

Mythology Photography **Fiction**
Fishing Christianity **Art** Cooking
Essays Buddhism Freemasonry
Medicine **Biology** Music **Ancient
Egypt** Evolution Carpentry Physics
Dance Geology **Mathematics** Fitness
Shakespeare **Folklore** Yoga Marketing
Confidence Immortality Biographies
Poetry **Psychology** Witchcraft
Electronics Chemistry History **Law**
Accounting **Philosophy** Anthropology
Alchemy Drama Quantum Mechanics
Atheism Sexual Health **Ancient History**
Entrepreneurship Languages Sport
Paleontology Needlework Islam
Metaphysics Investment Archaeology
Parenting Statistics Criminology
Motivational

PROGRESSIVE CARPENTRY.

Fifty Years' Experience in Building.

(During which time all known methods of construction
have been thoroughly studied.)

Many Valuable Improvements Made

WHICH ARE CLEARLY SET FORTH AND FULLY
EXPLAINED IN THIS WORK,

TOGETHER WITH

A SYSTEM OF FRAMING ROOFS

BY WHICH EVERY STICK CAN BE FITTED PERFECTLY WITH-
OUT MAKING EXTRA DRAWINGS TO OBTAIN CUTS,
BEVELS, ETC., FOR THE WORK.

———

By D. H. MELOY, ARCHITECT
(AND PRACTICAL BUILDER),
WATERBURY, CONN.

———

Fully Illustrated with Numerous Diagrams.

DAVID WILLIAMS COMPANY, Publishers,
232-238 William Street, New York.
1900.

PREFACE TO FIRST EDITION.

———

It has long been the purpose of the author to present to the public his new system of constructive carpentry—viz.: the plumb and level line system, by which all bevels and cuts may be obtained for any conceivable form of frame work, without the necessity of making extra drawing for that purpose, as required in other methods.

The work of preparation for this book was begun about thirty years ago, but the continued pressure of other business matters compelled deferment. It is now the sincere belief of the author that a good and wise providence controlled, because many very valuable and important improvements have recently been added, and which would not have appeared had the work been produced earlier, but are now presented.

It is also manifest to the author that this work is still incomplete, and that many valuable additions and improvements will soon be developed which will place it in the front rank as a reference book for all simple and reliable methods of construction, which is the author's highest ambition.

He contemplates soon to add to this book another part, treating upon other subjects invaluable to all who contemplate building. It is the earnest desire and hope of the author that he be able to continue the study of these methods and carry them on toward perfection.   Perfectness, however, in mechanics or mechanism will not be reached by any one so long as men aspire to higher attainments.   Therefore, any inquiry from the reader for more full explanations of any of the subjects, or any suggestions as to improvements, or even remarks from ubiquitous critics, will be most cordially welcomed, because they will all help to stimulate the author to his high purpose.

It is not alone for the little profit this book may possibly yield the author, but for the inestimable value it will be to thousands of young men who are seeking the best methods for doing their respective work.   But should both prove successful, the fullest anticipations of the author will be realized.

It is earnestly urged upon every young mechanic that he make himself familiar with all these methods.   He will then be worth one dollar per day more to his employer than without this knowledge, and can command higher wages for himself on account of this knowledge.

D. H. MELOY, ARCHITECT.

Waterbury, Conn., July, 1890.

CONTENTS.

A SYSTEM OF FRAMING ROOFS.

BY D. H. MELOY, ARCHITECT.

Before beginning to lay out the frame work of a roof it is necessary to know its size and form and what pitch the roof will have and dimensions of the timber to be used. All this is given in the plan, Fig. 1, which shows one-half of the roof. This building is 16 feet wide and 22 feet long. The roof will rise 6 inches to each foot, making it 4 feet high in the center. The timber to be used is of the following dimensions: Two wall plates, 4 x 8 inches, 22 feet long, and two wall plates 16 feet long, 4 x 8 inches; four hip rafters, 4 x 8 inches, 15 feet long; one ridge pole, 2 x 9 inches, 6 feet long; ten common rafters, 2 x 6 inches, 11 feet long; eight jack rafters, 2 x 6 inches, 8 feet 10 inches long; eight jack rafters, 2 x 6 inches, 6 feet 8 inches long; eight jack rafters, 2 x 6 inches, 4 feet 6 inches long; eight jack rafters, 2 x 6 inches, 2 feet 4 inches long. All the timber should be assorted and piled separately, so that you may know that it is full and what each stick is intended to make before any of it is laid out. Select straight sticks and of uniform size for making the pattern. No other plans or drawings of any kind are required to be made when this system of framing is once learned, and not a mark need be made more than what is marked on the timber to work to.

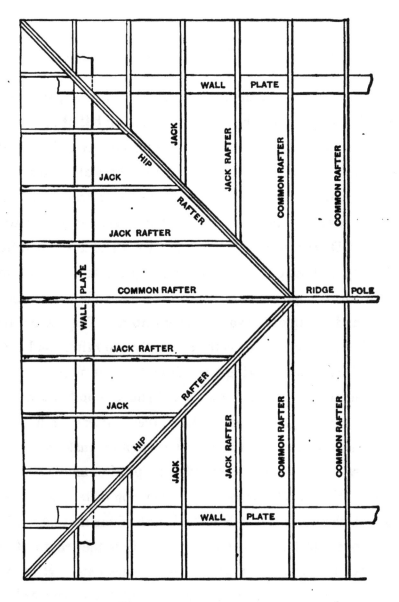

Fig. 1.—Plan of One End of a Hip Roof.

LENGTHS OF TIMBER IN HIP ROOFS.

Fig. 2 is another plan of the same roof as No. 1, and shows the method of obtaining the lengths for making bill of all the timber in this roof. Exact lengths of fitting are given further on. The lines AA represent the outer face of the wall plates. The lines BB represent the plan or run of the hip rafters. The lines CCCC represent the plan or run of the common and jack rafters. The line DD represents the pitch and length of the hip rafters (and is 15 feet long). The line EE represents the pitch and length of the common rafters. The lines FF, GG, HH represent the pitch and length of the various jack rafters. The line BE represents the ridge pole. BD is the rise of the hip rafters and BE is the rise of the common rafters, which is the same as the hips and is 4 feet high or rise in this roof. CF and CG and CH are the rise of the various jack rafters. All of these rafters can be measured by the scale 4 feet to 1 inch near enough for making a bill of the timber. D to D is the length of the hip rafter and is about 15 feet long. E to E is the length of the common rafter and is about 11 feet 3 inches long. F to F is the length of the first jack rafter and is about 8 feet 10 inches long. G to G is the length of the second jack rafter and is about 6 feet 8 inches long. H to H is the length of the third jack rafter and is about 4 feet 6 inches long. There will be short jacks at all the corners, which will be 2 feet 4 inches long.

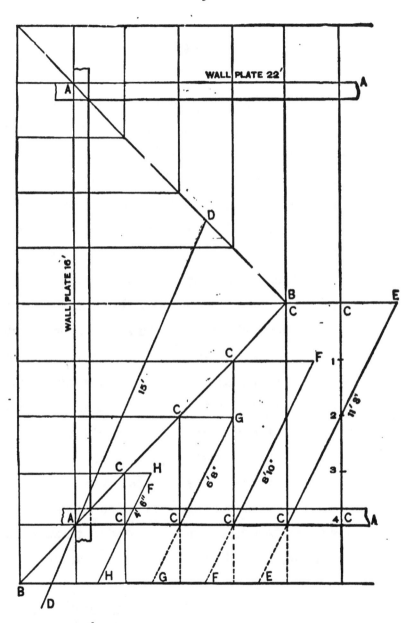

Fig. 2.—Method of Obtaining Lengths of Timber in a Hip Roof.

TO LAY OUT THE COMMON RAFTER.

Lay the square on the side of the stick to be fitted near the top end at figures 6 and 12 inches, the pitch of the roof, as shown in Fig. 3. Mark the down bevel A B. Square across the back of the rafter from A to C. Cut to marks C A B. To find the length of the common rafter: With a common pocket rule measure the distance on the steel square diagonal from 6 on the tongue to 12 inches on the blade, or from A to E, as shown in Fig. 3, and set it off along the top corner of the rafter as many times as there are feet in half the width of the roof, as shown by A E F. In this roof the distance on the steel square will be 13 and $7/16$ inches and must be set off eight times. Should there be a fractional part of a foot in the width of the roof, say 3 inches, place the square again on the stick same as before and at last point, and as for another run, and set off the distance required on the blade, as shown at G. Move the square along, keeping it on figures 6 and 12, and mark the down bevel H I through the point G. Always remember to take out the thickness of the ridge pole when laying out the rafters. Should there be a deck in the roof, the deck plates should be in line with the ridge pole, both on the face and on the top, so that the top end of all the rafters will fit alike on both the deck plates and the ridge poles, as shown by the sketch of deck roof.

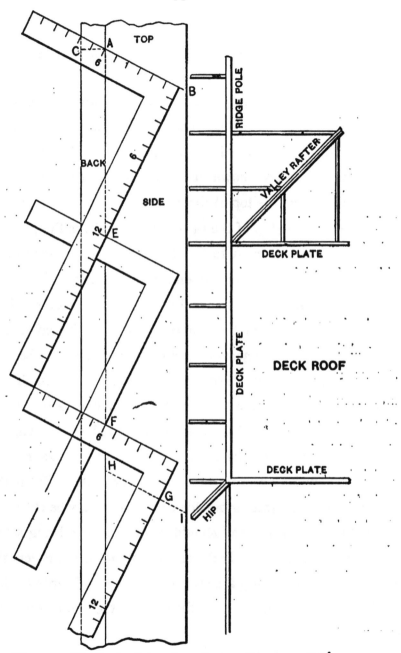

Fig. 3.—Method of Laying Out a Common Rafter.

TO LAY OUT SEAT OF COMMON RAFTERS.

Having found the length of the rafter according to and as shown in Fig. 3, and which point of length is at A, Fig. 4, lay the square on the side of the stick with the pitch of the roof 6 x 12 inches as before and as shown in Fig. 4, and mark the down bevel A B. Set off from the top of the stick on the down bevel line the height you wish the top of the rafter to stand above the plate, say 3 inches, as shown from A to C. Move the square along on the side of the stick toward the top end, until the edge of the blade comes to the point C, as shown by the square in dotted lines. Mark along the blade of the square the seat of the rafter C D. Extend the seat line full across the stick, whatever it may be, to D. Square across the underside or bottom of the rafter from B to E, and from D to F, and the lay out of the common rafter is completed. Cut to marks E B C, which is the plumb cut to fit against the outside face of the plate. Cut also to the marks F D C, which is the level cut to fit on the top of the plate. The bottom end of the rafter projecting outside of the plate may be cut to suit the detail of the cornice, and should be fitted before being set up, if any fitting is required. If the end is only to be cut to length it can be done best after they are set up, and the roof boards are laid.

Ridge pole should be set up the same as the seat of the rafter above the plate.

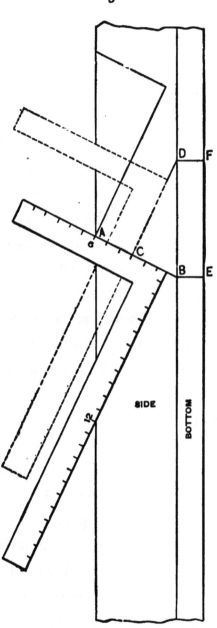

Fig. 4.—Method of Laying Out a Seat of Common Rafters.

TO LAY OUT THE JACK RAFTERS.

First determine how many jack rafters there will be on one side of the hip, which in this case is three rafters and four spaces, as shown in the plans, Figs. 1 and 2. Divide the length of the common rafter into that number of equal parts, which will give the different lengths of all the jack rafters, as shown at A and C, Fig. 5, which is one part, and shown also in the plans of the roof, Fig. 2, page 9, and marked 1, 2, 3, 4. The bottom ends of the jack rafters are laid out the same as the bottom ends of the common rafters, as shown in Fig. 4, page 13. The down bevel for the top of the jack rafters is also the same as the down bevel for the top of the common rafters, and is shown at A B, Fig. 5. To mark the side bevel, find the exact thickness of the stick at the end you are to fit and set off the distance on the side of the stick and at right angles from the down bevel to the top corner of the stick at E. Square across the back of the stick from E to F, and mark the diagonal A F, which will give the bevel required to fit against the side of the hips. Cut to marks F A B. Select a straight rafter for the pattern. When the patterns are accurately made according to the above rule, set bevels for laying out the other jacks, one for the plumb bevel and another for the side bevel. Fit jack rafters in pairs. The down bevel should be marked on both sides of the stick so as to cut the more accurately.

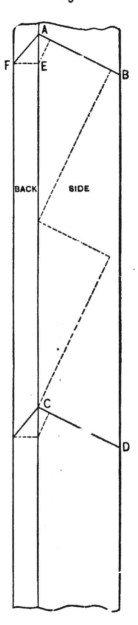

Fig. 5.—Method of Laying Out Jack Rafters.

TO LAY OUT HIP RAFTERS.

Hip rafters are laid out the same as common rafters, except that 17 inches is used for the run, instead of 12 inches as for the common rafter, because it requires 17 inches run for the hip rafter to equal 12 inches run for the common rafter, or because 17 inches is the hypothenuse of 12 inches square. Lay the square on the side of the stick at the top end, as shown in Fig. 6, with 6 inches and 17 inches the pitch of the hip rafter, and mark the down bevel A B. Find the exact thickness of the stick at this end, and set it off at right angles to the down bevel A B, and mark the top corner of the stick at C. Square across the back of the stick from C to the opposite side, E. Mark the diagonal A E. Cut to marks, E A B, which will give the side and down bevels against the ridge pole. If the hip rafter is to fit against the corner of a deck plate then square across the back from A to D, and lay out the opposite side the same as above, and mark diagonal, D C. Cut to marks A B, to the center F, and D B on opposite side of center F. If the hip is to fit against the straight ridge pole, cut through one way only, either right or left, as may be required, or as first laid out. This will complete the lay out of the top end of the hip rafter, and is shown in the cut as it will be after it is fitted.

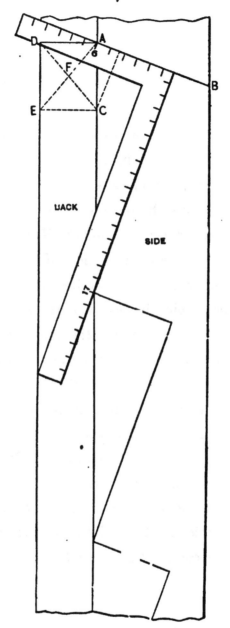

Fig. 6.—Method of Laying Out Hip Rafters.

TO FIND THE LENGTH OF HIP RAFTERS.

Take the distance with the common pocket rule, diagonal on the steel square, at figures 6 inches one way and 17 inches the other way—the pitch of the hip rafter, which, in this case, is 1 foot 6 inches and $\frac{1}{12}$ of an inch, and lay it off along the top corner of the stick, as shown in Fig. 7, from G to H and H to I. So continue as many times as there are feet in half the width of the roof, which, in this case, is 8 feet, so the lay off should be repeated eight times. The whole distance will be 12 feet and $\frac{1}{4}$ inch. The point of length is at I, and is the entire length of the hip rafter from outside of the plate to the center of the ridge pole. Always remember to take out half the thickness of the ridge pole, which in this case one-half the thickness is 1 inch, and remember, also, that for 1 inch of the ridge pole on the common rafter you must take out 1 $\frac{5}{12}$ inches on the hip rafter. So, then, move the square back 1 $\frac{5}{12}$ inches, as shown at J, and mark the down bevel J K through that point, which will give the exact length of the hip rafter, and in this case the entire length will be 11 feet 10¾ inches from outside corner of plates to the top corner of the ridge pole.

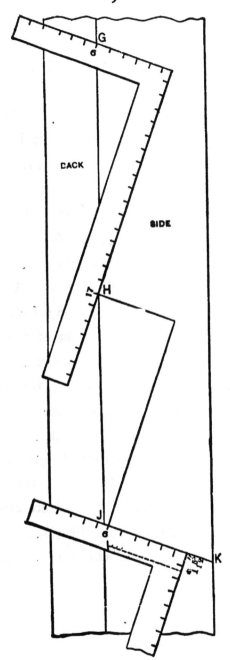

Fig. 7.—Finding the Length of Hip Rafters.

TO LAY OUT A FRACTIONAL PART OF A FOOT.

Should there be a fractional part of a foot in the width of the roof, place the square on the side of the stick below the last point made in the layout of the length, as shown at D E, Fig. 8, the same as for the run of another foot. Set off on the blade of the square the distance required, which, in this case, is 2 inches on the common rafter. Remember always to add $^1/_{12}$ of an inch to each inch used in laying out hips, because, as for 1 foot on the common rafter, 17 inches or 1 $^5/_{12}$ feet is required in laying out the hip rafter; so, also, for 1 inch on the common rafter, 1 $^5/_{12}$ inches is required in laying out the hip rafter. Therefore the distance to be set on the blade of the square for the extra 2 inches on the hip rafter is 2 $^{10}/_{12}$ inches and is shown at F. Now move the square down on the stick, keeping it at the same figures 6 and 17, as shown by the dotted lines, and mark the down bevel A B through the point F, which will give the exact length of the hip and the place of beginning to lay out the seat of the hip rafter. The rule for the layout is given in Fig. 9. Here again remember to take out one-half the thickness of the ridge pole before laying out the seat of the rafter, which will be 1 $^5/_{12}$ inches on the length of the hip rafter.

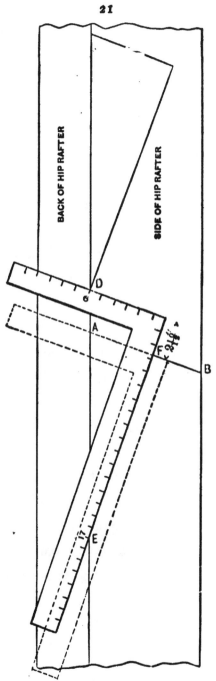

Fig. 8.—Method of Laying Out Fraction of a Foot.

TO LAY OUT LOWER END OF HIP RAFTER.

Having found the length of the hip rafter, which is at A, Fig.
9, lay the square on the side of the stick at that point with 6
inches and 17 inches—the pitch of the hip rafter, as shown—and
mark the down bevel A B. Set off from the top of the rafter at
A, on the down bevel line A B, the hight the hip rafter is to stand
above the plate, which, in this case, is 3 inches from A to C. Move
the square up so that the outer edge of the blade comes to the
point C and mark the seat line of the hip rafter along the blade of
the square through C. Find the distance across the top of the
plates where they lock together from outside to inside corners,
and where the hip rafter is to rest, setting it off on the seat line
from C to D. Make another down bevel line E F, through the
point D. Find the exact thickness of the hip rafter at this place and
set off the distance on the seat line from D to G and from C to H;
mark two more down bevels from G to I and from H to J. Square
across on the underside of the hip from F to F, I to I, B to B and
J to J, and lay out on the other side of the rafter the same as above.
Make the diagonal marks on the underside of the hip F I and B J,
both ways. Cut through all diagonal marks to the seat lines H
C G D, on both sides of the stick, and the layout of the hip rafter
is complete. Fit the lower end of the rafter outside of the plates
to suit the detail of the cornice.

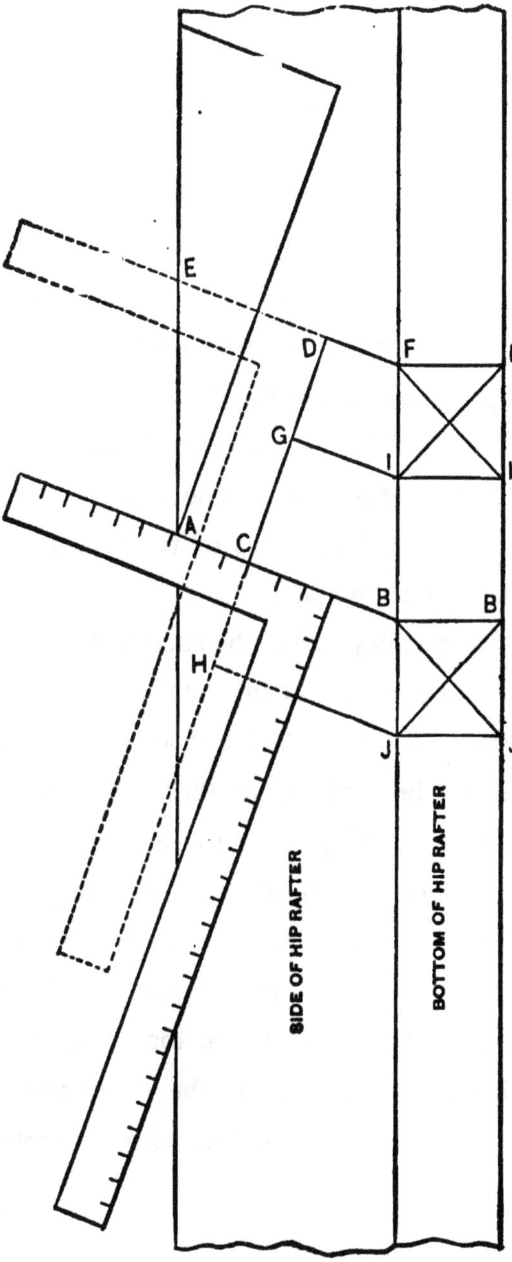

SIDE OF HIP RAFTER

BOTTOM OF HIP RAFTER

Fig. 9.—Laying Out the Lower End of a Hip Rafter.

TO LAY OUT BACKING OF HIP RAFTERS.

This rule is so simple that everybody wonders why it has been overlooked so long, or why it has not been used before in other works on framing rafters. It is easy to explain, and easy to understand. Take half the thickness of the hip rafter and set off from the top corner of the hip at right angles to any down bevel, as shown at A, Fig. 10. Or set off half the thickness of the hip on any level or seat line, as shown at C, Fig. 10. Gauge through these points on both sides of the hip, and bevel to the center on the back, as shown at B, Fig. 10. It is not very important that the hip and valley rafters be back beveled, and this extra work is often omitted even in many good buildings. When it is required, the bevel should be laid out and gauged on the sides of the stick before the seat of the hip is laid out, because the height of the hip above the plate is to be set off from the gauge line, instead of the corner of the stick. The valley rafter will also be set so much higher on the plate, and the jack rafters will be set even with the top of the valley if they are back beveled. When the valley rafter is not back beveled the top of the jack rafter must be set that much above the corners of the valley rafter, so that the top of the common rafter will be in line with the center and top of the valley rafter.

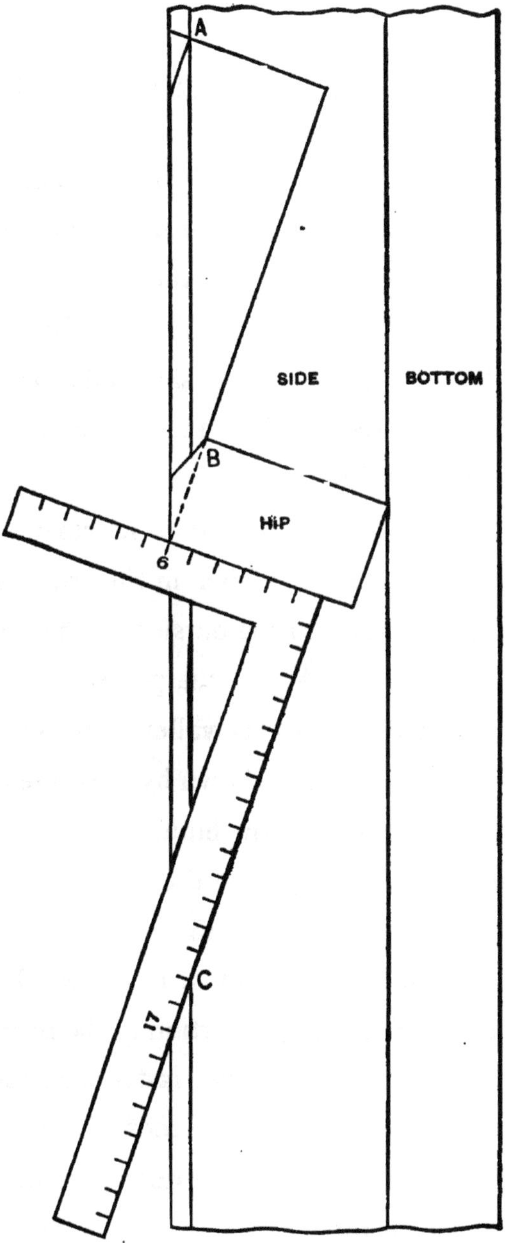

A

SIDE

BOTTOM

B

HiP

6

C

17

Fig. 10.—Backing a Hip Rafter.

TO INTERSECT ROOFS FOR TWO PITCHES

To lay out the rafters where roofs of different pitch intersect is much more difficult than in those of one common pitch, and is also much more difficult to illustrate and explain, therefore little has been said in other books on the subject. The roof we are to lay out has a pitch of 6 inches on one side and a pitch of 9 inches on the other side, as shown in Fig. 11, and the two roofs intersect in a valley shown by line A B.

Before laying out such a roof it is necessary to know the exact projection of the cornice, which, in this case, is 2 feet from the outside face of the plate to the outside of the crown molding on the cornice, as shown at A. If the projection of the cornice was only 1 foot, the lower end of the valley rafter would be nearer the angle of the building at C, as shown by the dotted lines C B.

If there were no cornice on the building the lower end of the valley rafter would be at the angle of the plates D, and the top of the plates would be even, but when there is a projecting cornice the plate on the 9-inch pitch roof must be raised above the plate on the 6-inch pitch roof just in proportion to the projection of the cornice, and in proportion as one rises faster than the other roof. In this roof the plate on the 9-inch pitch roof will be 6 inches above the plate on the 6-inch pitch roof, the difference of rise in the 2 feet projection of the cornice.

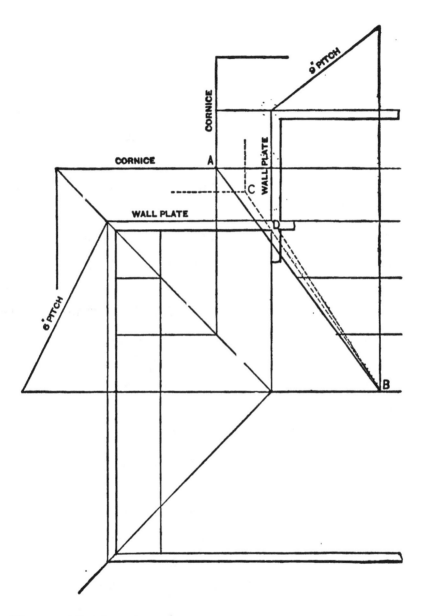

Fig. 11.—Laying Out Rafters for Roofs of Different Pitch.

TO LAY OUT THE VALLEY RAFTER FOR TWO PITCHES.

We have now found the position of the valley rafter, as described in Fig. 11 and shown in Fig. 12, by the line A B, which is the location and base line of the valley rafter and by which we can now obtain figures for laying out the work. Square in on the plan 1 foot from the outside face of the plate where the valley rafter is to stand from C to D. Mark D E parallel with the plate, cutting the valley rafter at E. From the point E set up the rise of 1 foot, which in this roof is 6 inches, from E to F, and mark the pitch line C F. By these lines, if correctly laid out full size, we obtain figures by which we can lay out the valley rafter. We will find the distance on the level or run from C to E to be just 15 inches, and this distance, C E, represents 1 foot run on this valley rafter the same as 17 inches represents 1 foot run in the common one pitch rafter. The rise from E to F is 6 inches and the pitch line from C to F will be 16 $\frac{3}{16}$ inches, so then we will lay out this valley rafter the same as the regular valley, using the figures 15 inches for the run, and 6 inches for the rise. The entire length of the valley rafter will be 8 feet 1⅛ inches or 16 $\frac{3}{16}$ inches six times. Remember to take out one-half the thickness of ridge pole. The side bevels of this valley rafter, and the side bevels of the jack rafter against the valley are given in Fig. 13.

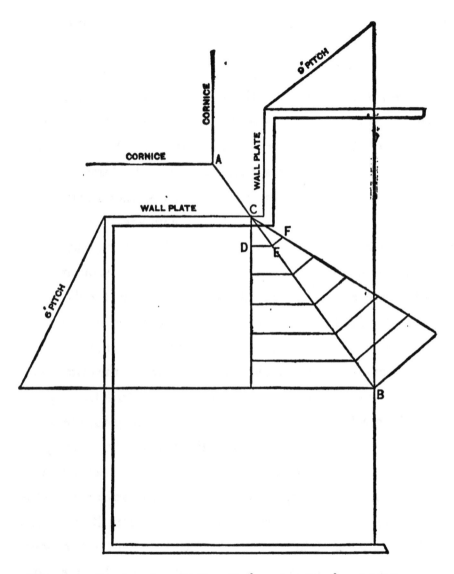

Fig. 12.—Laying Out Valley Rafter for Roofs of Different Pitches.

TO LAY OUT SIDE BEVELS FOR TWO PITCHES.

The line A B, of Fig. 13, represents the plan and position of the valley rafter. To find the side bevel of the valley rafter against the outside face of the plate, lay off on the plan the exact thickness of the rafter, parallel with the line A B, as shown by line C D; then square across from A to E, and the distance from E to D is the distance between the two down bevels on that side of ·the rafter. Mark the diagonal A D on the underside of the rafter, which will fit exactly against the outside face of the plate. The side bevels of the jack rafters are found by the same method. The line A F represents one side of the jack rafter for the 9-inch pitch roof. Find the exact thickness of the rafter and lay it off on the plan, parallel with A F, as shown by line G H; then square across from G to I, and the distance from A to I is the distance between the two down bevels on the jack rafter. Mark the diagonal A G, which will be the side bevel against the valley for the 9-inch pitch roof. The side bevel against the hip for the 6-inch pitch roof must be found by the same method, but the bevel will be longer. The line A J represents one side of the jack rafter and K L the other side or thickness. Square across from K to M and the distance from M to A is the distance between the two down bevels. Mark the diagonal A K and you have the side bevel against the valley for the 6-inch pitch roof.

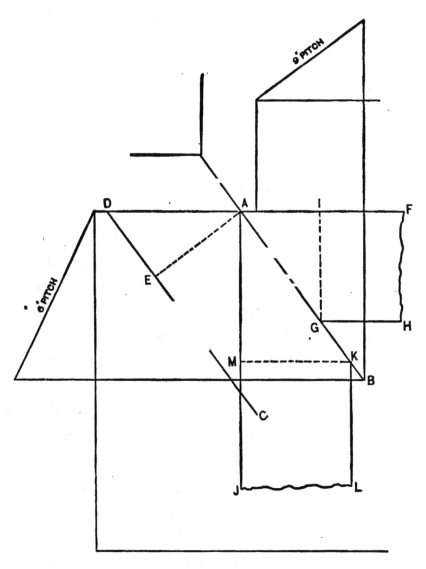

Fig. 13.—Laying Out Side Bevels for Two Pitches.

TO FIT PURLIN PLATES AGAINST HIP RAFTERS.

Place the square on the stick or on a board at figures 6 inches and 12 inches, which is the pitch of the roof, as shown in Fig. 14. Lay off the exact size of the top of the stick and the face sides of the stick both ways from the corner of the square B to A, 6 inches for the top side, and B to C, 6 inches for the face side, or the exact size of the stick, whatever it may be, either more or less. From these three points A B C square across the stick or board A to 1, B to 2, C to 3. Now square the end of the stick to be mitered on face D E, and on the top D F. Take the distance from 1 to 2 and set off on the bottom corner of the face, from E to G, and mark G D, which is the required bevel for the face of the purlin plate against the hip rafter. Take the distance 2 to 3, and set it off on the back corner of the top from F to H and mark H D, which is the required bevel for the top side of the purlin plate against the hip rafter. Cut to the marks G D H. The opposite sides of the stick may be marked by the same bevels, as shown by the dotted lines J G and J H, using the top bevel for the bottom, and the face bevel for the back. The bevels will be the same on any size stick having 6-inch pitch. Sticks of any size and pitch may be laid out by this rule, as is more clearly shown in Fig. 15.

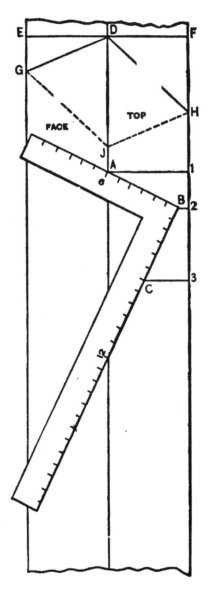

Fig. 14.—Fitting Purlins Against Hip Rafters.

TO MITER STICKS OF ANY SIZE AND PITCH.

Lay out the exact size and form of the stick in the position it is to be placed, as has been shown in Fig. 15. In Fig. 16 is represented a stick 6 x 8 inches, set to the pitch of 8 inches to 1 foot. Draw perpendicular lines from and touching all the corners. Now square the end of the stick to be mitered on the top A B and on face, A C. Take the distance from 1 to 2, Fig. 16, and set it off on the back corner of the top of the stick, from the square line B to D, and mark D A, which is the bevel required on the top side of the stick. Then take the distance from 3 to 4, and set it off on the bottom corner of the face of the stick from C to E. Mark E A, which is the bevel required on the face side of the stick. Cut to marks E A D. The opposite sides of the stick may be marked by the same bevels, as shown by the dotted lines F E and F D. Fig. 17 represents a stick same size as that shown in Fig. 16, but is set to a pitch of 12 inches to 1 foot. The bevels for Fig. 17 are laid out by the same rule as in Fig. 16. Mark the perpendicular lines from all the corners, and take the distance from 1 to 2 for the bevel on the top of the stick, and the distance from 3 to 4 for the bevel on the face of the stick. Cut to marks J K L and the dotted lines M on the opposite side of the stick. If the stick to be mitered is not square, lay it out as it is and as shown at Fig. 17.

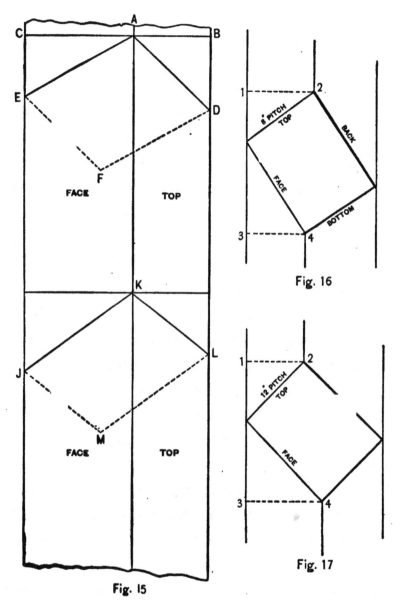

Fig. 15

Fig. 16

Fig. 17

Figs. 15, 16 and 17.—Method of Mitering Timber of any Size and Pitch.

TO MITER ROOF BOARDS AT HIPS AND VALLEYS.

Lay out a section of the board full size and in the position that it is to be placed on the roof, 6 inches to 1 foot, as shown in Fig. 18, A B showing the width and face of the board and B C the thickness. Draw a perpendicular line from the upper corner of the face from B to D. Draw level lines from the lower corner of the face from A to D and from the under corner from C to E. The distance from A to D, Fig. 18, is the bevel required for the face or width of the board, and the distance from C to E is the bevel for the edge or thickness of the board. The bevels on the hips and in the valley will be alike, except the bevels on the hips will be shortest on the upper edge and on the under side of the boards, but in the valleys the bevels will be the reverse—they will be longest on the upper edge and on the under side. The bevels for the roof boards at hips and valleys on any pitch roof may be found by this same rule. Fig. 19 shows how to obtain the bevels of the board on a roof having 12-inch pitch. A B shows the width of the board and B C the thickness of the board. Draw the plumb line B D and the level lines A D and C E, as before described and shown in Fig. 18. The distance from A D in Fig. 19 is the bevel for the face of the board, and the distance from C to E is the bevel for the edge or thickness of the board.

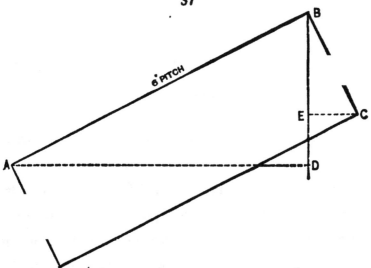

Fig. 18.—*Mitering Roof Boards at Hips and Valleys.*

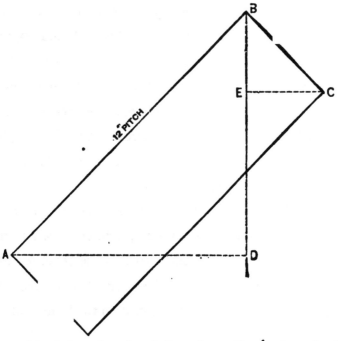

Fig. 19.—*Obtaining Bevels of Board on Roof of 12-Inch Pitch.*

TO MITER BOARDS IN VALLEY OF TWO PITCHES.

Lay out sections of the boards to be used full size, whatever the width or thickness may be, in the position they are to be laid on the roofs, as shown in Fig. 20. A B C D represents a section of the covering board having 6-inch pitch, and A B C D of Fig. 21 represents a section of the covering board having 8-inch pitch. Draw level lines from the top face corner of the boards from B to E. Draw a plumb line touching the lower face corner of the boards at A from E to F. The layout of the covering boards so far in Figs. 20 and 21 is the same as in Figs. 18 and 19, and will give the bevels for mitering the covering board in the valley, having one common pitch. But to find the bevels required for mitering the covering board in the valley where two roofs of different pitch intersect, proceed as follows: Draw a line 8-inch pitch, as shown from A to G, Fig. 20. The distance from G to E is the bevel for the face of the board on the 6-inch pitch roof, and the distance from D to F is the bevel for thickness of the board on the 6-inch pitch roof. Make a line 6-inch pitch, as shown from A to G in Fig. 21. The distance from G to E is the bevel for the face of the board on the 8-inch pitch roof, and the distance from D to F is the bevel for the thickness of the board on the 8-inch pitch roof.

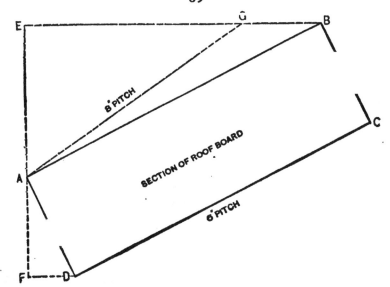

Fig. 20.—Position of Roof Board Having 6-Inch Pitch.

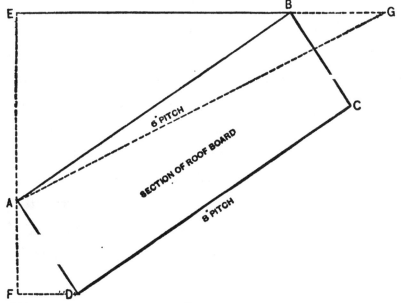

Fig. 21.—Position of Roof Board Having 8-Inch Pitch.

PUTTING TOGETHER ROOF FRAME-WORK.

It is just as important to know how to put together the frame-work of a roof properly as it is to know how to lay it out correctly. First know that the wall plates are straight on the outside face and on the top side. If the plates rest on brick or stone walls and are not uniform in height they should be wedged up under side until the top is perfectly straight. Roofing slate of various thickness, or points of shingle, make good wedges, and should be placed directly under where the rafters are to rest on the plates. When the plates are firmly secured to the walls with the proper wall anchors all the interstices between the wedges should be underpointed with mortar. The next thing to do is to set up the deck plates or ridge poles on stanchions of exact hight. Then set up the hip rafters, and if they are fitted correctly they will bring the deck or ridge pole into its proper position, but keep the stanchions under until the roof is covered. When the hip rafters are secured at the bottom and top see that they are straight all ways, and that the sides of the hips are plumb. Brace and stay the hip rafters straight and plumb before setting the jack rafters, then all the jacks will fit them nicely and hold them straight. Stanchions should remain under valley rafters for all time.

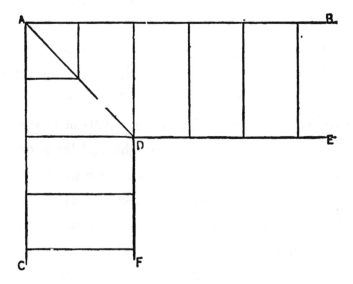

Fig. 22.—Plan of Corner of Roof with Curved Rafters.

TO FRAME CURVED ROOFS.

The plan, Fig. 22, represents the corner portion of a roof with curved rafters. The lines A B and A C represent the outside face of the main wall plates, and the lines D E and D F represent the outside face of the deck plates, which is 10 feet above the main plate, and is 4 feet back from the face of main plate, as shown in Fig. 23. This figure represents the common rafter and the curve of the roof, which is 1 foot, and the radius 15 feet, but the curve may be made more or less as desired. To make the pattern for the common rafter in this roof, take a board about 15 inches wide and 12 feet long; make one edge straight, and lay it out on the straight edge of the board the same as for a common rafter in a straight roof, with 10 feet rise and 4 feet run, as shown in Fig. 23, the line A B representing the edge of the board and the pitch of the roof. Divide the length of the rafter into any number of equal parts, as shown and numbered 1, 2, 3, and so on. Mark level lines across the face of the board from all the points, 1, 2, 3, and so on. Then make the curved line from A to B as desired, which in this case is 15 feet radius or 1 foot curve. The layout of the common rafter pattern is now complete, but before cutting it we must lay out the hip rafter, because it is necessary to use the lines and figures in the common rafter for the layout of the hip rafter.

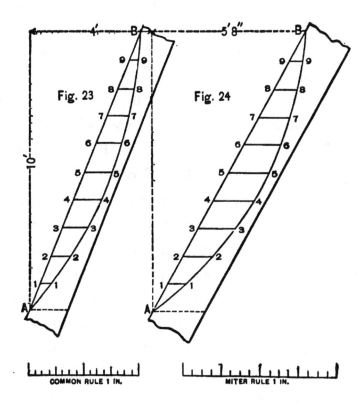

Figs. 23 and 24.—Laying Out Curved Rafter.

TO LAY OUT CURVED HIPS.

Take another board, about 18 inches wide and 12 feet long, and make the edge straight. Lay out the rise and run of the hip rafter on the straight edge of the board, the same as for a straight hip rafter, but remember to use 17 inches for the run in the hip rafter for each foot of run in the common rafter, so that the run of the hip will be four times 17 inches, which will be 5 feet 8 inches, as shown in Fig. 24. The line A B of the figure is the pitch and length of the hip rafter. Divide the distance from A to B, Fig. 24, into the same number of equal parts as the line A B, Fig. 23, and draw level lines from each point across the face of the board, 1, 2, 3, and so on, as shown in Fig. 24. We must now use what I call the miter rule, shown in Fig. 25, and which is described further on. Make all the distances on the hip rafter 1 1, 2 2, 3 3, and so on, of Fig. 24, equal to the distances on the common rafter, Fig. 23, measuring the common rafter with the common rule, but lay out the hip rafter, Fig. 24, with the miter rule. Mark the curved line through all the points 1, 2, 3, and so on. The back level is found by measuring back on all the level lines half the thickness of the hip rafter, and bevel from the center, or by moving the pattern back half the thickness of the hip and marking the bevel line on both sides and beveling to the center.

TO MAKE THE MITER AND OCTAGON RULE.

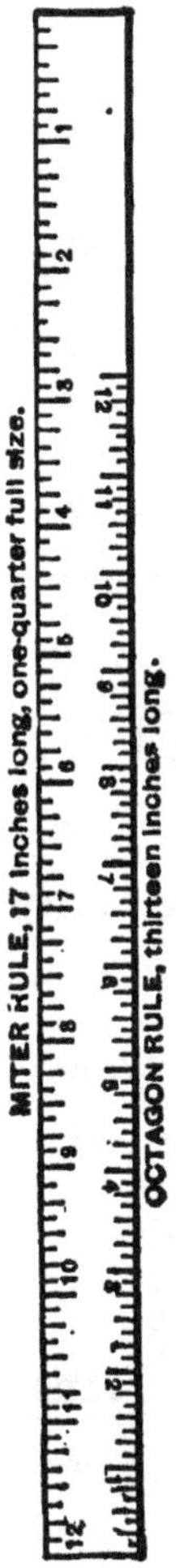

MITER RULE, 17 inches long, one-quarter full size.

OCTAGON RULE, thirteen inches long.

Fig. 25.—Miter and Octagon Rule.

Take a strip of straight grained wood about ¼ inch thick and ½ inch wide and 17 inches long. Divide the distance, 17 inches, into 12 equal parts, as shown in Fig. 24, then divide each of these 12 parts into eight equal parts, the same as on the common rule. The whole length, 17 inches, represents 1 foot on the miter line. Each of 12 parts represents 1 inch on the miter line and each of the eight parts represents one-eighth of 1 inch on the miter line. The length of any brace having an equal run each way can be obtained with this miter rule. If the run of a brace be 2 feet 3 inches each way, the length of the brace will be 2 feet 3 inches, measuring with the miter rule. The octagon rule is made the same as the miter rule. Lay off on the opposite side of the miter rule 13 inches, which represents 1 foot on the run of an octagon hip. Divide the 13 inches into 12 equal parts and the 12 parts into eight equal parts. This rule will lay out hip rafters for octagon roofs, the same as you lay out hip rafters for a square roof with the miter rule.

TO FRAME OCTAGON CURVED ROOFS.

Fig. 26 represents the base or wall plates of an octagon roof with double curved rafters, the lower half being convex in form and the upper half concave, as shown in Fig. 27. This figure represents the common rafter and is in the middle of each of the eight straight sides from A to the center, as shown in Fig. 26. In Fig. 27 are represented the hip rafters which are to stand on the eight angles from B to the center, as shown in Fig. 26. To make the pattern for the common rafter, take a board of sufficient size and make one edge straight, mark the form of the curves as desired, but always have two points somewhere in the form come to the straight edge of the pattern board so as to govern the pitch or level lines for forming the hip rafter. Having made the pattern for the common rafter as above, or as desired, divide the entire length from C to E into any number of equal parts and mark level lines across the board from all the points, as shown in Fig. 27, and numbered E 2, 3 and so on. Take another board for the hip rafter pattern, lay it out to the proper length and pitch, divide it into the same number of equal parts as the common rafter and mark the level lines, as shown and numbered E, 2, 3, as in Fig. 28. Transfer all the distance in the common rafter pattern E, 1, 2 2, 3 3, and so on, to the hip rafter pattern. Measure the common rafter with the common rule, but measure the hip rafter with the octagon rule.

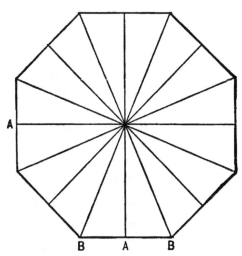

Fig. 26.—Plan of Octagon Roof With Double Curved Rafters.

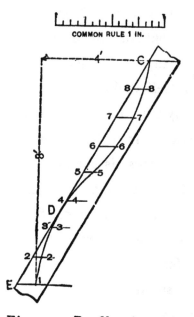

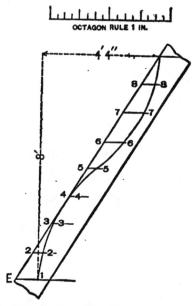

Fig. 27.—Profile of Common Rafter.

Fig. 28.—Profile of Hip Rafter.

NEW CONSTRUCTION OF A TRUSS ROOF.

Many forms of self-supporting or truss roofs have been given in nearly all books published on Carpentry, and many of them are very good and well constructed for certain kinds of buildings, but the one here represented, and illustrated by Fig. 29, has never been given in any other book.   This form of truss is designed for manufacturing buildings, storage buildings, &c., or wherever room saved is a consideration.   Since its introduction in 1856 it has been largely used for such buildings, wherever it has become known to builders.   These trusses are so constructed that the floor room under the roof is just as good and useful for storage purposes as any other room in the building, there being no obstructions, except a few iron rods that support the floor below. This form of truss is also very rigid in itself, and also braces the lateral sway of the building more than any other form of truss can possibly do, because the main brace extends from floor beam to roof beam, and the thrust pressure of the truss is put upon the floor beam, instead of the side walls of the building.   The more the side walls of the building are extended above the upper floor, as shown by Fig. 30, the more rigid the whole roof will become, which is a very important consideration for buildings where heavy machinery is to be run at a high rate of speed.

When a roof of wider span is required, this same form of truss can be extended and constructed as represented by Fig. 31.

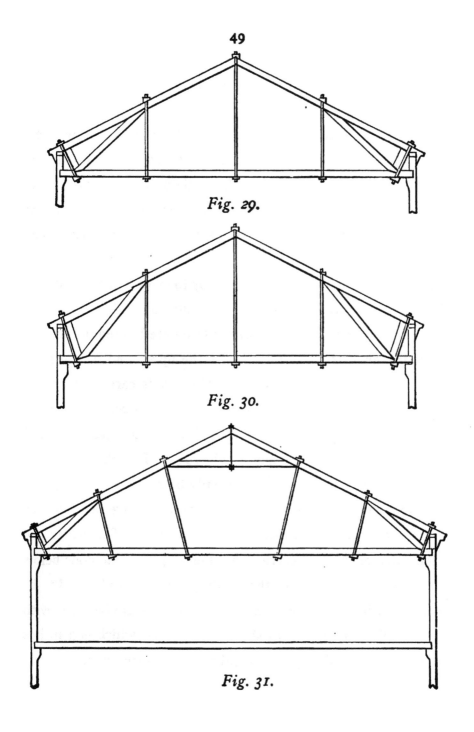

Fig. 29.

Fig. 30.

Fig. 31.

DETAILS OF A TRUSS ROOF.

Fig. 32 represents a detail section of this truss, drawn on a larger scale of ¼ inch to one foot, to show more clearly the dimensions of the timber required and the method of framing the truss. For a building 30 to 40 feet wide the timber should be as follows, and should be of yellow pine: Main floor beams, 8 x 12 inches. Main truss rafters, 8 x 10 inches. Strut brace, 8 x 8 inches. The length of the timber will be governed by the width of the building, which any practical mechanic can readily determine. The two main supporting rods in this form of truss are located in the middle of the main truss rafters and at the upper end of the strut brace, and marked A A, Fig. 32. These rods carry the greater weight of the roof, and should be 1¼ inches diameter in a roof of 30 feet span, and 1⅜ inches in a roof of 35 feet span, and 1½ inches in a roof of 40 feet span. The rods B B at the foot of the main rafters and strut braces are simply tie rods binding together the whole truss, and carry but little strain, and are not required to be more than ¾ inch diameter for any span. The center rod at the ridge of the roof carries only the weight of the floor below and its load in the center of the building, and is not required to be more than 1 inch diameter. This form of truss can be made with very low pitch if desired, as shown by Fig. 33, which is 3 inches to the foot, and will be just as rigid as with 6-inch pitch.

· B

1¼ in. ROD

Fig. 32.

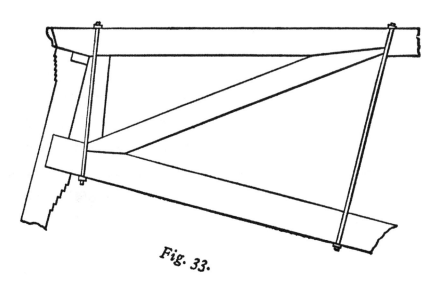

Fig. 33.

Since the publication of this book I have received such a multitude of letters from Carpenters all over the country in relation to frame-work of various kinds, and especially for truss roofs for large buildings and churches and for bridges, that I have been constrained to add a few pages on such work. The forms of trusses that I now present have all been built by myself, and have been thoroughly tested and have proven very satisfactory, being both simple of construction and very substantial and can be relied upon when properly constructed of good material. When we have accomplished results that are perfectly satisfactory, it is a waste of time and material to experiment upon plans and theories that have not been put to tests and thoroughly demonstrated as to merit. A great many forms of trusses have been made and published as possessing meritorious points, but many of them are very cumbersome in their construction, and (many of them) often contain a large amount of unnecessary material and therefore a large amount of extra work in framing them, besides adding extra weight to the roof, which is a very serious objection. (Lightness and strength are the two important things in truss roofs.) Other forms have been designed and published, and in their illustrations and descriptions have a good show, but when constructed and put to test have proved practically worthless and a failure from some unseen weak points.

I have often been called to examine defective roofs and bridges, therefore have been compelled, from necessity, to give much time and thought to the investigation of such matters, and have many times been astonished at the manifest lack of judgment and capability in the construction of much of the work that I have inspected. I have often been inquired of also as to the best material to be used in this class of work. I have used very many kinds of timber, and have found them all suitable for roof and truss work, viz.: White pine, yellow pine, cypress, spruce, chestnut and oak. In my experience I have found white pine timber to be best for a truss roof, because it is light. It will not spring, wind or crack so much as other timber. It is brought together and will remain so much better and longer than other timber. I made many very large roofs of white pine, more than 50 years ago, that appear to be as good now as when they were first built, and I see no reason why they will not be a hundred years more. Cypress and yellow pine timber would be my next choice for a good truss roof, and I think cypress better than yellow pine, because of its lightness and toughness. Yellow pine timber is much heavier than either pine or cypress, but for the same roof yellow pine may be one or two inches in size smaller than either, because it is much stronger than either pine or cypress.

THE CHEAPEST AND BEST TRUSS FOR ROOFS.

Among the many forms of trusses that have been presented in other publications there is none better and more simple of construction than the two here illustrated by Figs. 34 and 35. The one constructed like Fig. 34 is the more simple, and is more generally used on buildings from 30 to 40 feet in width.

For small buildings the timber may be quite light, and boards strongly nailed on the sides of the beams and rafters may be substituted for rods and bolts, but for good permanent buildings I should use rods, and timber of good size and quality. It is not wise to use too small timber in a truss roof. For a roof thirty feet wide I should make the truss beam 8 x 12 inches. The truss rafter 8 x 10 inches. The brace 6 x 8 inches. The center rod 1¼ inches. Two short rods 1 inch. Foot bolts ¾ inch.

Fig. 35 represents a similar truss, but calculated for buildings from 40 to 60 feet wide, and can be extended by adding rods and braces even to 80 feet or more, but the timber must be larger in proportion as the span of the roof is increased. I should make the material for a truss to span sixty feet about as follows: Beams, 10 x 14 inches. Rafter, 10 x 12 inches. Long brace, 6 x 10 inches. Short brace, 4 x 10 inches. Rods, 1½ inches, 1¼ inches, 1 inch and ¾ inch. Any carpenter capable of framing a truss roof can calculate the proper size material for the roof he is to frame.

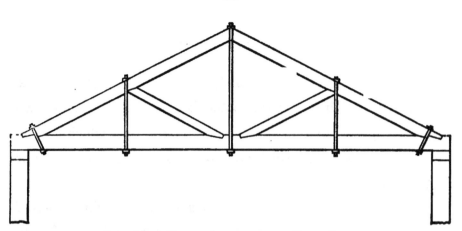

Fig. 34.—Truss for 30 to 40 Foot Span.

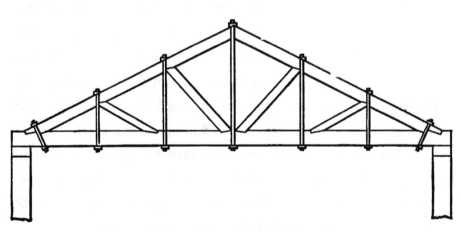

Fig. 35.—Truss for 40 to 60 Foot Span.

TRUSSES FOR 50 TO 80 FOOT SPAN.

Many trusses have been made with King and Queen posts, held together with iron straps, bolts, &c., but I discarded all such trusses long ago, and adopted more simple forms, which are decidedly better, some of which are now illustrated and described by Figs. 34 to 37.

If I were employed to construct a wood truss that required very much iron work, as straps, saddles, stirrups, bands, &c., to fasten it together, I should recommend that it be made entirely of iron, because it would cost but little more than if made of wood. Besides, iron construction is fast becoming a leading factor in the building industry of the country, especially for large roofs and bridges, and I consider them decidedly better for permanent work. The two forms of wood trusses shown on the opposite page, Figs. 36 and 37, are varied a little from the two forms given before, Figs. 34 and 35. In these trusses I have set the lightning rods obliquely and the braces parallel with and against the rods and fitted in square between the truss beam and the truss rafters, so that when the nuts are tightened the work is brought together very rigid. I have represented straining beams at the top of these trusses, because they are calculated for greater span than Figs. 34 and 35, and to prevent splicing the main truss rafters. I would not use straining beams when truss rafters can be got long enough to reach the full span without splicing.

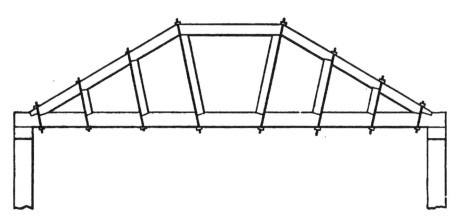

Fig. 36.—Truss for 50-Foot Span.

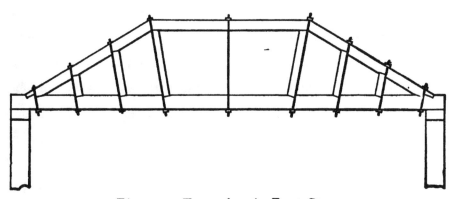

Fig. 37.—Truss for 60-Foot Span.

TRUSSES FROM 80 TO 100 FOOT SPAN.

The two forms of trusses, Figs. 38 and 39, are intended for large roofs of wide span, from 80 to 100 feet. The principle of the trusses is the same as the first, Figs. 34 and 35, the only difference being the addition of more sections or rods and braces sufficient for the increased width of the buildings they are to span. These trusses are suitable for large churches or halls, or for large mills and store houses, or any buildings that require self supporting roofs. The material for these large trusses should be increased in size according to the increase of span.

For a span of 80 feet like Fig. 38, I should make the truss beam 12 x 15 inches, the truss rafters 12 x 12 inches, straining beam, 10 x 12 inches. Long strut braces, 8 x 12 inches. Short braces, 6 x 12 inches. Two main rods at top of main rafters, 1⅜ inches. Two next rods, 1⅛ inches. Two next rods and center rod, ⅞ inch. Four foot bolts, ¾ inch. Purlin plates, 4 x 8 inches, just below each rod. Common rafters, 2 x 6 inches, 20 inches between centers. Ridge pole, 2 x 8 inches. Roof trusses may be placed 8 feet to 12 feet apart, as suits the layout of the buildings best. This truss is sufficiently strong to carry the next floor below if it is so desired, but in such case the two main rods in the truss should be 1⅝ inches, and the rods that support the floor beams below should be 1⅜ inches.

All rods should be upset and swaged at the ends, so that they will not be decreased in size by cutting threads.

Fig. 38.—Truss for 80-Foot Span.

Fig. 39.—Truss for 100-Foot Span.

TRUSSES FOR LOW PITCH ROOFS.

The two trusses represented on opposite page, Figs. 40 and 41, are calculated for roofs of low pitch and wide span, and where metal, gravel, or ready roofing is to be used for covering These roofs pitch 3 inches to one foot, and are as low pitch as self supporting truss roofs should be made if the span is over 50 feet.

Fig. 40 is constructed upon the same principle as the two on previous page, but without the straining beam, and to show also that such roofs may be extended to any desired width or span by splicing the tie beams and main rafters.

Fig. 41 is constructed entirely different, and is intended for large freight houses and sheds. It will be seen that there is but one iron rod in this truss and the foot bolts at the bottom of the main truss rafters. All the other supports being boards about 10 inches wide, strongly nailed on the sides of the tie beam, and the main truss rafters as indicated in the sketch, Fig. 41. The tie beam and the truss rafters should be 4 x 12 inches, neatly fitted together at the top and bottom, and the bolts tightened up so as to give the tie beam the proper camber desired. Then the board ties and braces are to be nailed on, placing the first about 10 feet below the ridge, and at right angles with the truss rafter, the next 9 feet below, and the next 8 feet, and so on to the bottom. Then fit the braces in between each tie, as shown in the cut, Fig. 41, and strongly nail all.

Fig. 40

Fig. 41

Truss for Low Pitch Roof.

TRUSSES FOR ROAD BRIDGES.

For small bridges there is no form of truss better than those I have given for truss roofs, but the timber should be much larger for bridge trusses. If a bridge is to be large there is no better truss than that known as the Howe truss.

If the bridge is to be housed in or covered (and it is better that all bridge trusses should be covered) there is nothing better than white pine timber for the main truss, and it is very important that all the joints be well coated with thick paint when it is put together, and the joints kept well filled with white lead which will add many years to the life of the truss.

Fig. 42 represents a truss suitable for a small bridge of about 20 or 25 feet span.

Fig. 43 represents a truss suitable for a bridge of 30 or 40 feet span.

Fig. 44 represents a truss suitable for a larger bridge, from 40 to 50 feet span. The main string beams should be about 15 x 20 inches. Main braces and straining beams 12 x 15 inches. Small braces 8 x 12 inches. Floor beams and wall sills 15 x 18 inches. Main rods 1⅝ inches. Shorter rods 1⅜ inches. Foot bolts 1⅛ inches. Floor joist 4 x 10 inches, about 3 feet apart. Floor plank 2½ inches thick, or more if desired, and from 6 to 8 inches in width, make the best floor.

Fig. 44 may be extended to 60 feet, by making the center section double width, as in Fig. 43.

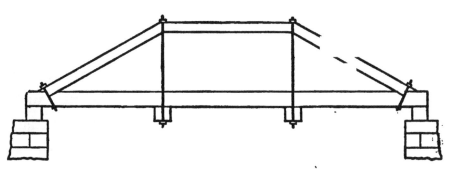

Fig. 42.—Truss for 30-Foot Span.

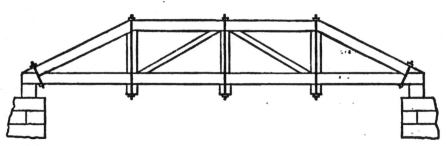

Fig. 43.—Truss for 40-Foot Span.

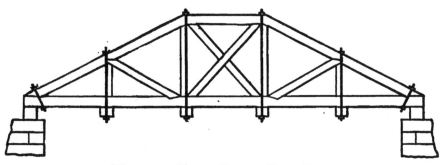

Fig. 44.—Truss for 50-Foot Span.

TRUSS ROOFS FOR SMALL CHURCHES OR HALLS.

It is always desirable in the construction of small church roofs to utilize as much space as possible above the main walls or plates, and at the same time make the roof self supporting. Fig. 45 represents a roof of such construction, and one that has been extensively used and for a long time. I consider it the best among the many that has been presented, yet it has one weak point. I have always found that the ridge will sag in the middle, and the side walls or plates will round out the same. I have often been called to examine roofs of this kind, and have always found them sagged and spread as stated. To prevent this sagging and spreading I have recommended the addition of two iron rods running through the top of the main truss rafter and the tie beam where the main brace crosses the tie beam, which is shown in Fig. 45 by dotted lines marked "iron rod."

Fig. 46 represents another construction of truss that I have found to be very rigid for a small roof, and utilizes all the space under the roof if it is so desired; besides, the timber can all be left exposed and planed and ornamented to taste.

The timber for a roof 30 or 40 feet wide should be as follows: Main rafters, 8 x 10 inches; collar beams, 8 x 8 inches; wall plates, 6 x 8 inches; main posts, 8 x 8 inches. Angle braces, 8 inches thick, same as the timber, and to be bolted through the main timber, as shown in Fig. 46.

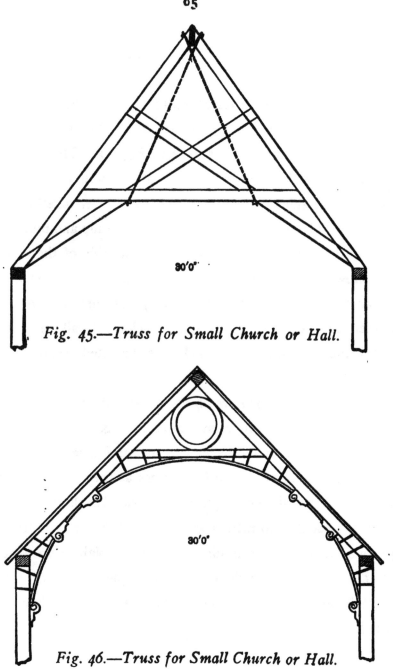

Fig. 45.—*Truss for Small Church or Hall.*

Fig. 46.—*Truss for Small Church or Hall.*

FRAMING HIP ROOFS THAT ARE NOT SQUARE.

It has been the universal custom to set the jack rafters parallel with the side walls or plates of the building, where the building is not square, as is shown in Fig. 47. There are many reasons why such a custom is wrong. If you make the jack rafters on the sides and ends of such roofs meet together on the hips they will be much nearer together on one hip and much wider apart on the other hip than they are on the side of the building, as is shown in Fig. 47. If you set the jack rafters on the end the same distance apart as the side jacks they will not meet together on the hips. Besides the jacks will all be different lengths and bevels, and each jack must be fitted separately to fit its own place.

Now, if the jack rafters on the bevel end of the roof are set at right angles to the plate, as shown in Fig. 48, each pair will be of the same length and bevel, and the same distance apart, and all will meet together on the hip rafters as they should; besides, the roof boards will all cut square. The side bevels for these jack rafters against the hip rafters will be found the same as for roofs with two pitches, which is illustrated and explained on pages 30 and 31, Fig. 13, and which is the most simple and correct method for obtaining bevels of jack rafters that are not set at right angles to the hips.

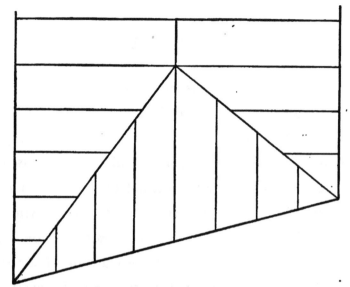

Fig. 47.—Hip Roofs that Are Not Square—Old Way.

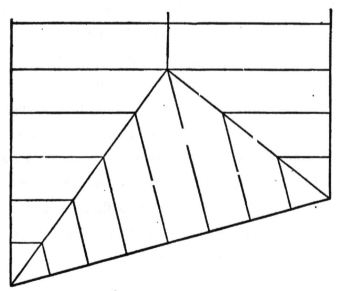

Fig. 48.—Hip Roofs that Are Not Square— New Way.

TRUSSING SADDLE ROOFS.

Small buildings for barns and storehouses are often built one and one-half story high, and the side plates in such case are not more than 5 or 6 feet above the upper floor. To prevent such a building from spreading in the middle iron rods are put through the plates from side to side, but these iron rods are considered very objectionable and troublesome when moving goods around in the building, and it has been often asked how the roof can be supported and the iron rods dispensed with. The construction of such roofs is shown in Fig. 49. Board braces, say 1 by 6 inches, may be set from the foot of the two end rafters on the plates to the top of the center rafter at ridge pole. Let the braces into all the rafters flush with the top of the rafters and nail them all together strongly before putting on the roof boards. Small roofs can be held straight by putting all the roof boards on diagonal, same as the braces, and nailing them strongly to all the rafters, or by strongly spiking the braces on the under side of the rafters. Another way is to truss the ridge pole, as shown in Fig. 50. If the ridge pole is kept straight, and the rafters strongly secured to it, the roof cannot sag or the building spread. Be sure that the plates are kept straight until the stays and braces are all strongly secured and the roof completed.

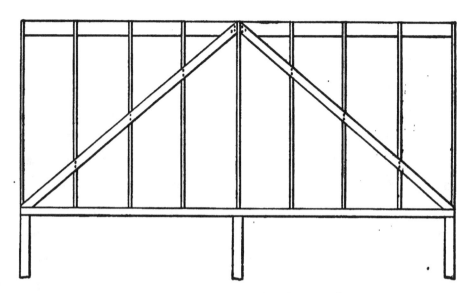

Fig. 49.—Bracing Saddle Roof.

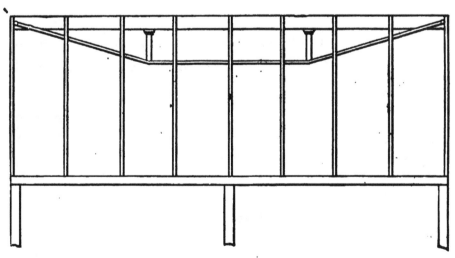

Fig. 50.—Bracing Saddle Roof.

PITCH AND LENGTH OF RAFTERS.

The pitch of rafters is always understood to mean the number of inches they rise in one foot run, and is shown in Fig. 51. The length of rafters is, therefore, determined by measuring with the pocket rule diagonally across the steel square from 12 inches on the blade for the run to whatever the rise is on the tongue, which will give the length of rafter for each foot of run. These distances are all given on the opposite page from 1-inch pitch to 24-inch pitch, in Fig. 51. Multiply these distances by the number of feet there is in half the width of the building, and you will have the length of the common rafters; or mark that distance off on the top corner of the rafter the same number of times. The blade of the square will mark the bottom or level cut, and the tongue will mark the top or plumb cut. Any pitch rafter can be laid out by the same method.

Hip and valley rafters can be laid out the same as common rafters, only use 17 inches for the run of hips and valleys instead of 12 inches.

If the hip rafters are not back beveled, the hip rafter must be set just so much lower on the plate as the becking would cut off from the corners.

If the valley rafters are not back beveled, the top of jack rafters must be set that much above the top corners of the valleys so that the tops of the jack rafters will range to the center of the valley rafers.

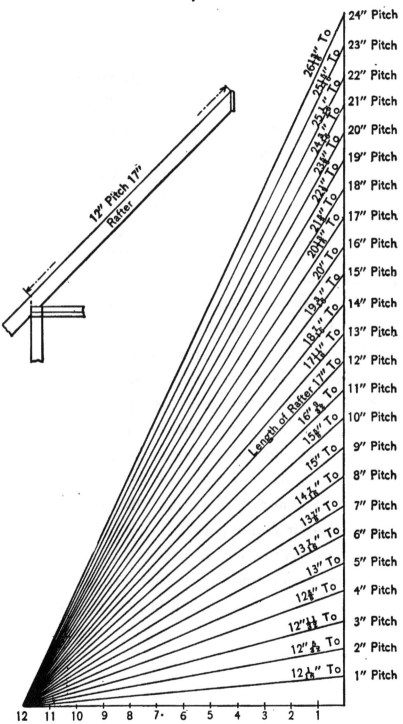

LENGTHS AND BEVELS OF JACK RAFTERS.

The best and quickest way to obtain the lengths of jack rafters has been given on pages 14 and 15, Fig. 5. Many other methods have been given by other authors, but there is none so simple and accurate as this. First determine how many jacks there will be on the hips or valleys, then divide the length of the first or longest jack rafter into as many equal parts as there will be spaces between the jacks, which will give the number and length of each jack rafter, as has been previously described on page 14 and illustrated on page 15, Fig. 5.

The best method for obtaining the side bevels of jack rafters against the hips and valleys in every kind of roof is also clearly given on pages 30 and 31, Fig. 13, which is simple and accurate.

Another good method for marking the side bevels of jacks is with the steel square, and shown on the opposite page. Take the length of one foot of the rafter on the blade of the square, which for 12-inch pitch is 17 inches, and lay it on the back of the rafter at the top of the down bevel, and one foot run on the tongue of the square, and mark by the blade, which will give the bevel against the hip for 12-inch pitch.

All side bevels for jack rafters from 1-inch pitch to 24-inch pitch are also given on the opposite page, Fig. 52.

Bevels can be set by those lines and relied upon as being correct.

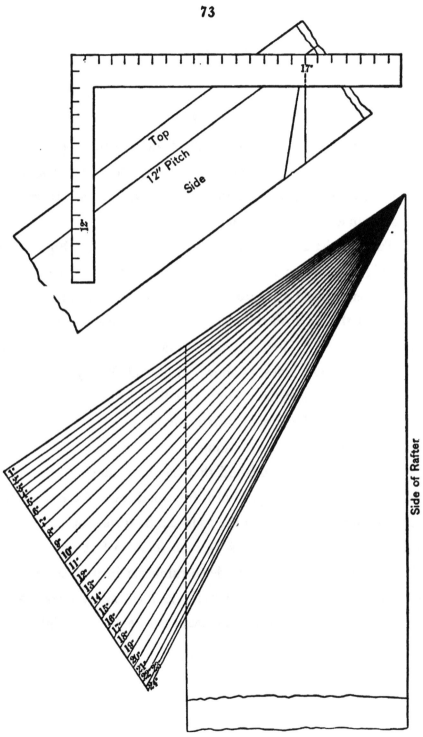

Top

12" Pitch

Side

12

17"

Side of Rafter

WRONG METHOD OF TRUSSING LONG JOIST.

Inquiry is often made, " How can long joist be trussed so as to avoid center beams and columns ? "

Two methods are given and illustrated on the opposite page that have been recommended and have often been used, but always proved worthless. I give my experience with them, and any one can easily try the same.

Take a joist 2 x 12 inches and 30 feet long, like Fig. 53, set it on edge, as it would be in the building. Draw a line on the top from end to end and measure how much it sags in the middle. Then turn the joist on the side, and make it camber 1 inch, or more if you like. Then nail on the brace strips as strong as you please, and on both sides, as shown in the cut, Fig. 53. Then set it on edge again as at first, and you will find that it sags just the same as before. Then don't waste any time and material on this method, for it is good for nothing.

Fig. 55 is another method that has been much used, and is good for iron joist, and good also for wood joist, if the construction is such that the nuts on the rods can be turned up as the timber shrinks. I have seen 2 x 15 inch joist 30 feet long trussed with rods like Fig. 55, and made to camber 1 inch when first set, but after two years the joist shrank ½ inch, and the rods were loose and of no use because they could not be got at to tighten up the nuts.

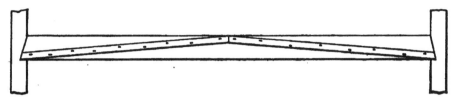

Fig. 53.—Wrong Method of Trussing Long Joist.

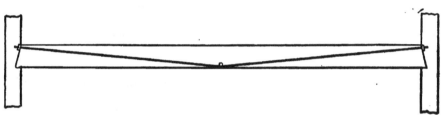

Fig. 55.—Wrong Method of Trussing Long Joist.

Plan of Floor.

RIGHT METHOD OF TRUSSING LONG JOIST.

On the opposite page two methods of trussing long joist are given, both of which have proved to be very effectual and simple of construction.

Fig. 56 illustrates the construction of a floor with joist 2 x 12 inches and 30 feet long, having ¾-inch truss rods between the joist in each alternate space, and with one center supporting timber, 4 x 6 inches, continuous under all the joist, as shown in the cut, Fig. 56.

Fig. 57 is the same method as the above, but illustrates double supporting beams under the joist which are 2 x 12 inches 40 feet long. The rods for this span should be about 1⅛ inches and ⅞ inch for 35-foot span. I don't think it necessary to increase the size of the joist for any width floor that is trussed by this method.

The joist can be fitted and placed in the building the same as other joist that are not trussed, and the rods can be put in any time before the underside is prepared for lathing or sheathing.

The beams underside of the joist should be put in place on stanchions when the joist are placed in the building, and the stanchions should remain until the rods are put in. The timber on the top and at the ends of the joist should be yellow pine, about 3 x 4 inches, notched into the joist 3 inches, so that the top will be even with the lining floor. The base in the room should be put on so that it can be removed to turn up the nuts on the rods when the joist shrink.

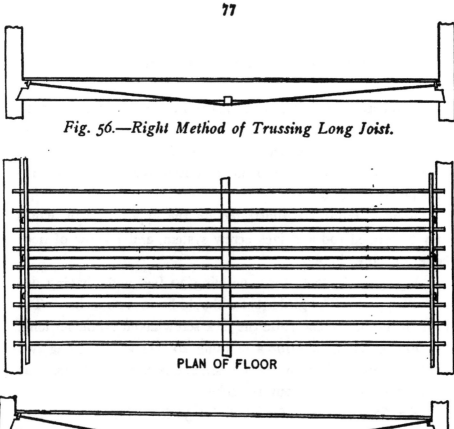

Fig. 56.—Right Method of Trussing Long Joist.

PLAN OF FLOOR

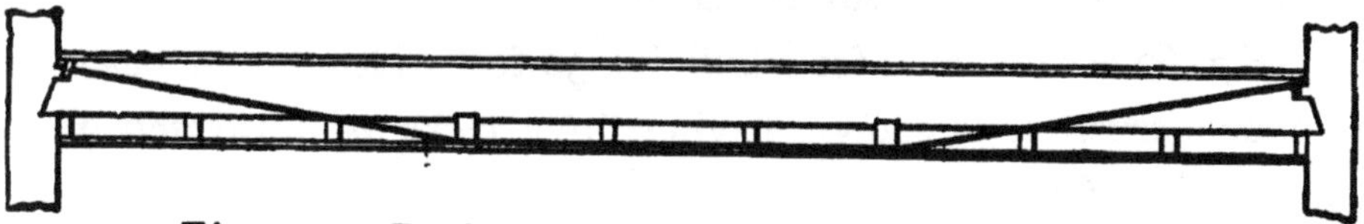

Fig. 57.—Right Method of Trussing Long Joist.

PLAN OF FLOOR

METHOD OF TRUSSING GIRDER BEAMS.

Inquiries have often been made how floor beams can be trussed so as to dispense with columns or posts, which many times are considered very objectionable; especially where it is desirable to have the room free from all obstructions, as many business places demand.

Four methods are illustrated on the opposite page, which may be termed rod and jack screw system, and I know of none better or more simple.

Fig. 58 is suitable for a building 30 feet wide, having one center support. The girder beam should be 10 x 12 inches. The rods should be 2 inches, and the jack screw about 20 inches long.

Fig. 59 is suitable for a building 40 feet wide, having two straining jacks, as shown. The beam and rods should be same as above, and the jacks about 18 inches.

Fig. 60 is calculated for a building 50 feet wide, and the beam is double trussed, as shown, having three jacks as illustrated in the cut, Fig. 60.

This beam should be about 12 x 15 inches, and the main long rod 2¼ inches. The short rods 1¾ inches. The center jack 2 feet long, and the small jacks 1 foot long.

Fig. 61 represents a beam for a building 60 feet wide, having a line of columns in the center and trussed on each side, as shown by the cut, Fig. 61.

This beam should be 12 x 15 inches, and the rod on each end 2 inches in diameter. The straining jacks about 20 inches long.

Fig. 58.—Method of Trussing Floor Beam—30-Foot Span.

Fig. 59.—Method of Trussing Floor Beam—40-Foot Span.

Fig. 60.—Method of Trussing Floor Beam—50-Foot Span.

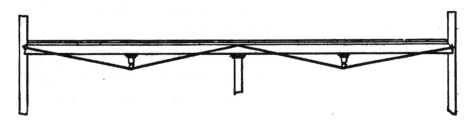

Fig. 61.—Method of Trussing Floor Beam—60-Foot Span.

METHODS OF SPLICING TRUSS BEAMS.

If timber can be procured long enough to span the full width of building it is better to do so, especially for tie beams in a truss roof, because roof beams require to be spliced much stronger than floor beams. If roof tie beams are over 50 feet long it becomes necessary to splice them or build them up with planks bolted together, which I think better than splicing.

Figs. 62, 63 and 64 on the opposite page illustrate three methods for splicing truss beams, and are good enough for any building, and either of which are suitable for truss beams, 8 x 10 inches to 12 x 15 inches, and from 60 to 100 feet long, and are also illustrated in Figs. 38 and 40.

When splicing or framing truss beams avoid cutting into the timber any more than is absolutely necessary to make suitable seats or shoulders for the braces or ties.

Fig. 62 is a butt joint splice fastened together with another piece of timber 10 x 12 inches, notched into the underside 3 inches, and another piece, 3 x 12 inches, on the top, and all to be double, bolted together, as shown in the cut, Fig. 62.

Figs. 63 and 64 are spliced by lapping together, as shown, and to be strengthened with other pieces of timber all bolted together, same as above, and as shown in the cuts.

Floor beams do not require to be spliced together so strongly as roof beams. They may butt together and fasten with joint bolts or dog hooks, or they may lap and bolt, as shown in the cuts, Figs. 64 and 65.

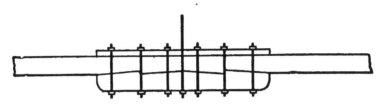

Fig. 62.—Method of Splicing Truss Beam.

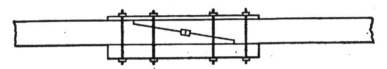

Fig. 63.—Method of Splicing Truss Beam.

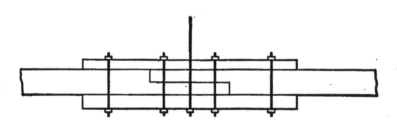

Fig. 64.—Method of Splicing Truss Beam.

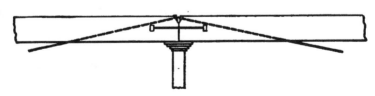

Fig. 65.—Method of Splicing Truss Beam.

MITERING FASCIA SET WITH PITCH OF ROOF.

Lay out a section of the fascia on the pitch of the roof, say 2 inches for the width and ⅞ inch for the thickness, or whatever it is, as shown in Fig. 66, A B being the width and A C the thickness. Draw plumb lines from the corners of the face B D and thickness C E, as shown. The distance from A D is the bevel on the face for 2 inches of the width, and the distance from A E is the bevel for the thickness. This will give the bevels for any width fascia set to 6-inch pitch. The same rule will give the bevels for any width fascia set to any other pitch, if the section is laid out on the right pitch. Fig. 67 shows how to mark the bevels on the fascia. Square the end of the fascia on the face A B and edge A C, make a short gauge line on the face 2 inches from the bottom edge, as shown by the dotted line B C, make the distance on the gauge line B C, Fig. 67, equal to the A D, Fig. 66, and the distance C D on the back corner of the edge equal to A E, Fig. 66. Mark from the square line on the lower face corner, A, through C on the gauge line the full width of the fascia, which will give the bevel for the face. Mark from the corner A, Fig. 67, to D, which will give the bevel for the edge or thickness. The lines A D and A C give the miter joint.

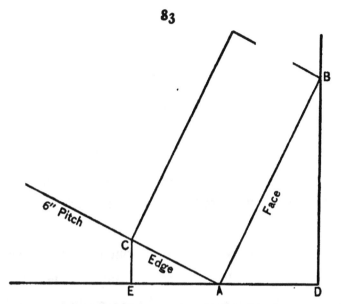

Fig. 66.—Mitering Fascia Set With Pitch of Roof.

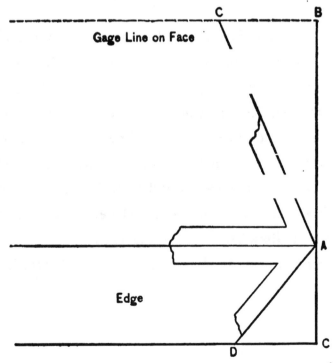

'Fig. 67.—Mitering Fascia Set With Pitch of Roof.

MITER BEVELS FOR FASCIA OF ANY WIDTH OR PITCH.

The rule for finding the bevels to miter fascia when set with the pitch of the roof, as given on the previous page and shown in Figs. 66 and 67, will apply to any other pitch and to fascia of any width.

If the section of the fascia is laid out full size the bevels will be more accurate, but they will be the same whether the fascia is 2, 4, or 6 inches wide, or more. I have made the width of the fascia in the cuts only 2 inches for want of room to make them wider.

In Fig. 68 are given all the bevels for mitering fascia set to pitch of roof from 1 to 24 inch pitch, both for the face and the edge, and may be relied upon as being correct. Bevels may be taken from these lines, placing the stock of the bevel on the full level line A B, and move the blade of the bevel on the bevel line marked to the pitch of the roof 4, 6, 8, or whatever the pitch may be. The bevel lines above the full level line A B are the bevels for the face or width of the fascia, whatever the width may be.

The bevel lines below the full level line A B are the bevels for the edge or thickness of the fascia, whatever the thickness may be.

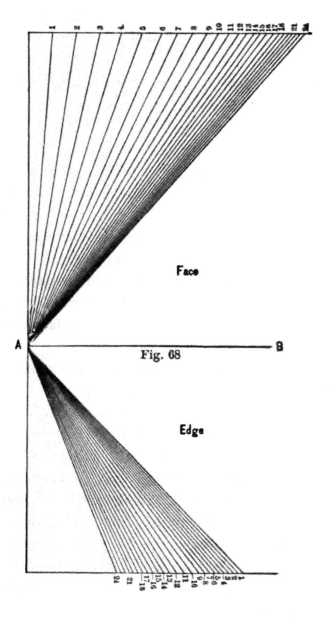

Fig. 68.—Miter Bevels for Fascia of Any Pitch.

METHOD OF OBTAINING BEVELS FOR ANY PITCH ROOF.

On page 87, Fig. 69, is another sketch showing how to obtain bevels for mitering fascia set to pitch of roof 12 inches to 1 foot instead of 6 inches, as in Fig. 66, page 84, and is given to show that bevels are obtained by the same rule for any pitch. The distance from A to B, Fig. 69, is the distance to bevel the face for 2 inches in width, and the distance from A to C is the distance to bevel the edge or thickness.

In Fig. 70 are given all the distances for beveling fascia from 1-inch pitch to 24-inch pitch. The long radiating lines from A to 1, 2, 3, etc., on the left, Fig. 70, represent the pitch of the roof from 1 to 24 inches. The radiating lines from A to the figures 1, 2, 3, etc., on the right, Fig. 70, represent the face of the fascia, and all are at right angles to the pitch lines.

The circular line B D represents the width, though it may be any width desired, the bevel will be the same for the same pitch. The distances from the full plumb line A D, Fig. 70, to the figures 1, 2, 3, etc., on the curved line B D are the distance to bevel the fascia for each pitch—viz., D to 1 for 1-inch pitch, D to 2 for 2-inch pitch, D to 3 for 3-inch pitch, and so on to 24-inch pitch. The circular line C E represents the thickness of the fascia, which is usually ⅞ inch thick, but may be any other thickness without changing the bevels. The distance from the plumb line A D to the intersection of the circular line C E, and the various pitch lines are the distances to bevel the edge or thickness of the fascia.

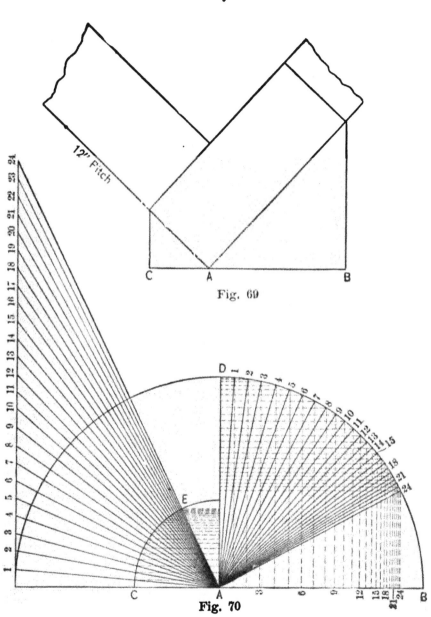

Fig. 69

Fig. 70

Method of Obtaining Bevels for Any Pitch.

MITERING PLANCEER SET TO PITCH OF ROOF.

The rule for finding the bevels for mitering planceer set with the pitch of the roof is the same as for mitering fascia, the only difference being the bevels are the reverse. Two sections are given on the opposite page, Figs. 71 and 72, showing the difference between 6-inch pitch and 12-inch pitch.

The distance from A B is the distance to bevel the face for 3 inches wide, and the distance from A C is the distance to bevel the edge or thickness ⅞ inch.

The bevels for mitering the planceer are the same as the bevels for mitering the fascia, and are all given in Fig. 68, page 88, from 1 to 24 inches. The bevel for the face of the planceer is the same as the bevel for the edge or thickness of the fascia, and the bevel for the thickness of the planceer is the same as the bevel for the face of the fascia. Bevels may be taken for both planceer and fascia, and for any pitch roof from 1 to 24 inches, from the bevel lines given in Fig. 68, page 88, and may be relied upon as being correct. No matter what the width or thickness of the stuff may be, the miter bevels given in Fig. 68 will always be the same for the same pitch, but the bevels for the planceer will be the reverse of the bevels for the fascia. The bevels for the inner angles will also be the reverse of the outer corners, so that the workman must use his own judgment in applying the bevels to the work.

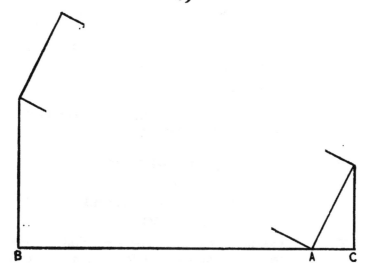

Fig. 71.—*Mitering Planceer Set With the Pitch of Roof.*

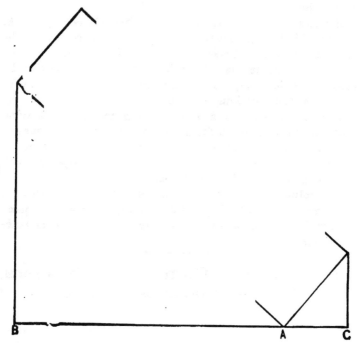

Fig. 72.- -*Mitering Planceer Set With the Pitch of Roof.*

Hicks's Builders' Guide.

Comprising an Easy, Practical System of Estimating Materials and Labor for Carpenters, Contractors and Builders.

A comprehensive guide to those engaged in the various branches of the building trades.

By I. P. HICKS.

160 Pages, 114 Illustrations.   Bound In Cloth, Price $1.00.

This book presents to the trade, first, a practical system of estimating materials, in concise form for reference.   In the next place is given the average day's work of all kinds, and the average rates on which to figure in almost all details of building construction.

Material and Labor, including Lumber, Carpentry Work, Masonry, Plastering, Hardware, Painting, etc., are among the principal divisions. These in turn are illustrated and ably treated under numerous subdivisions, showing step by step the method of estimating, contracting and building, together with practical methods of doing the work.

This is followed by a section given to the Geometrical Measurements of Roof Surfaces, with numerous illustrations and examples.   It embraces an easy system of framing curved roofs, and simple methods of roof framing of all descriptions, with practical examples and illustrations.

The volume also contains a chapter on Mitering Planceers, Mouldings, etc., illustrating and describing the making of many joints which are the source of no little trouble and annoyance to workmen.

David Williams Company, Publishers,
232-238 William Street, New York.

CPSIA information can be obtained
at www.ICGtesting.com
Printed in the USA
BVHW061243180219
540527BV00025B/2026/P

9 781330 029978